Ecosystems of North America

The Atlantic Coast

Olivia Skelton

BENCHMARK BOOKS
MARSHALL CAVENDISH
NEW YORK

Series Consultant: John T. Tanacredi, Ph.D., Supervisory Ecologist, National Park Service, U.S. Department of the Interior

Consultant: Richard Haley, Director, Goodwin Conservation Center

Benchmark Books
Marshall Cavendish Corporation
99 White Plains Road
Tarrytown, New York 10591-9001

Library of Congress Cataloging-in-Publication Data

Skelton, Olivia.
The Atlantic coast / Olivia Skelton.
p. cm.—(Ecosystems of North America)
Includes bibliographical references and index.
Summary: Examines the tides, plants, animals, and ecosystems found along the Atlantic coast.
ISBN 0-7614-0903-3 (lib. bdg.)
1. Coastal ecology—Atlantic Coast (U.S.)—Juvenile literature. [1. Coastal ecology—Atlantic Coast (U.S.) 2. Ecology—Atlantic Coast (U.S.)] I. Title II. Series.
QH104.5.A84S54 2000 98-48645
577.5'1'0974—dc21 CIP
AC

Photo Credits

The photographs in this book are used by permission and through the courtesy of:
Animals Animals/Earth Scenes: Ted Levin 9, 42-43; Herb Segars 22; Michael Gadomski 24-25; Wendy Neefus 29; Zig Leszczynski 37, 53; Patti Murray 44, back cover. ***Photo Researchers, Inc:*** Jeff Lepore front cover; Adam Jones 14-15; Andrew J. Martinez 16, 19; Rafael Macia 32; Michael Gadomski 38; S. R. Maglione 45; Gilbert S. Grant 46. ***Tom Stack & Associates:*** Sharon Gerig 4-5, 12; Tom and Therisa Stack 8, 51; Brian Parker 18, 50, 54; Tom Stack 20, 58; John Shaw 26; David and Tess Young 28; Matt Bradley 34-35; Larry Lipsky 48-49; W. Perry Conway 56-57; David Young 59. Cover design by Ann Antoshak for BBI.

Series Created and Produced by BOOK BUILDERS INCORPORATED

Printed in Hong Kong
6 5 4 3 2

Contents

Exploring the Atlantic Coast

The Atlantic coast stretches more than 2,000 miles along the eastern edge of North America, from Maine to Florida. If you traveled from one end to the other, you would see many different landscapes: rocky cliffs, sandy beaches, wide bays, grassy marshes, and sunken reefs. In some places, forests reach the water's edge; in others, grassy fields or sand dunes stretch along the shore. There are large cities on the Atlantic coast, such as New York, Boston, and Miami. But there are also small towns, isolated seashores, and nature preserves.

Change is constant on the Atlantic coast. Storms roar in from the ocean. They crumble cliffs, tear up beaches, and cut new waterways through sandy islands. Old marshes fill in as new ones appear. Tides rise, fall, and rise again.

Would you like to explore the Atlantic coast? Start in Maine, where you might walk along a rocky beach at the base of jagged cliffs. You will need to close your jacket tightly against the cold ocean wind and sea spray. At your feet, you might see a small pool of water holding red starfish, seaweed, tiny crabs, and sea anemones.

The Portland Head Lighthouse has guarded the rocky Maine coast since 1791.

In North Carolina, you can walk for miles along wide, sandy beaches. Take off your shoes on a hot day and feel the cool ocean waves wash over your feet and between your toes. Those holes in the wet sand reveal the hiding places of lugworms and clams. Leave the water's edge and walk inland, behind the sand dunes. There, you might find a shady forest of holly, pine, and sassafras trees.

On the broad Chesapeake Bay, look for fleets of fishing boats. The wooden boats with white sails are skipjacks—the only fleet of sailboats that still fishes in American waters. For hundreds of years, fishers in boats such as these have scooped blue crabs and oysters from the bay's waters.

Pull on heavy boots to tramp through a Georgia salt marsh. Peer through the tall blades of cordgrass standing at eye level all around. The buzz of insects fills the air. Herons dart overhead as snowy egrets wade along the muddy shore. Fiddler crabs creep through the shallow water, searching for food.

Continuing south, you dive into Florida's warm, blue-green waters and swim with bright blue parrotfish and golden Spanish hogfish. Delicate coral formations are all around you, shaped like fans, antlers, and feathers.

As you can see, millions of plants and animals live along the Atlantic coast, in many different environments. But are they just scattered in a random way? What determines where and how they live?

Many Types of Communities

Deer, crabs, sugar maples, moray eels, starfish, wild cranberries, moose, and gulls. These are just some of the **organisms**, or living things, found along the Atlantic coast. But moray eels do not live in Maine, and cranberries do not grow in Florida. Why not?

The reason is simple. Each organism requires a specific type of habitat in order to live and grow. A **habitat** is a place that has all of the living and nonliving things an organism needs to survive. Organisms live where the food, climate, soil, and sunlight suit them best. So the warm coastal waters of Florida are the perfect habitat for moray eels, just as the cool climate and acidic soils of coastal New England are just right for cranberries.

The Atlantic Coast

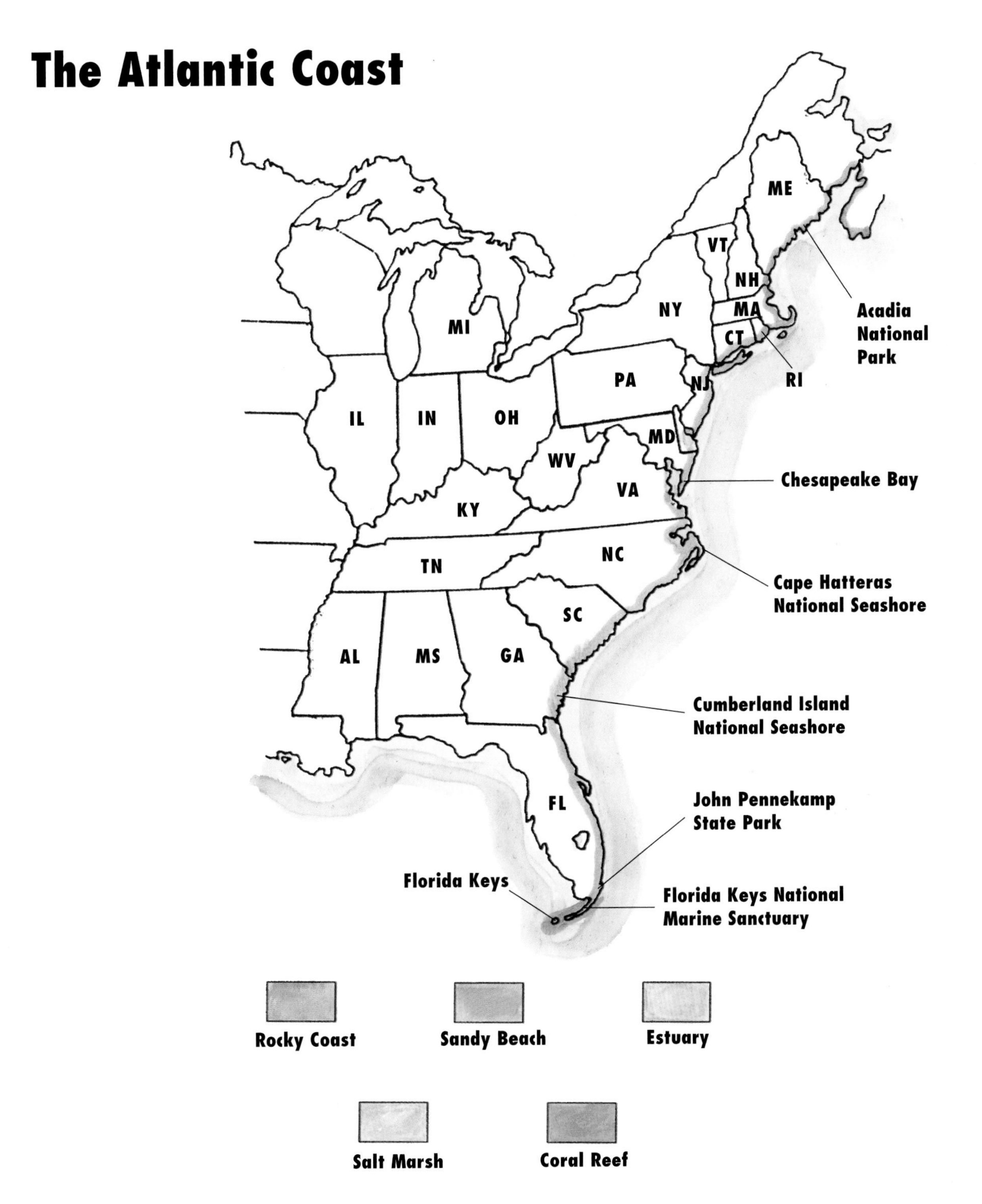

The Atlantic coast changes gradually from a rocky shoreline to a coral reef. It also includes the unique communities of salt marshes and estuaries.

The moray eel lives in the shallow waters of coral reefs and rocky coasts.

Many animals and plants thrive in the same general conditions. They do not all eat the same foods, need the same soil, or live in exactly the same spot. But they share a common place, interacting in ways that strongly affect each other. Together they form a **biological community**, a group of organisms that live together and interact in a particular environment.

A number of organisms—such as eelgrass, beetles, seagulls, clams, and raccoons—might form a small part of the biological community on a sandy beach. Many of these organisms depend on one another for food or protection. For example, seagulls eat clams, and raccoons often eat the eggs of seagulls. Grasses growing on sand dunes high on the beach might provide food for insects or shelter for birds and small animals. Take away the grass or remove the seagull and the entire community is affected.

But organisms in a community do not interact only with one another. They also interact with the nonliving things around them in larger groupings called ecosystems. An **ecosystem** is the association of living things in a biological community, plus their interaction with the nonliving parts of the environment. The beach sand and ocean water are part of the ecosystem. The ecosystem also includes the climate, the sunlight shining on the beach—even chemical pollution that could harm the biological community. So no organism exists on its own. The moray eel, the cranberry plant, all are part of a vast array of interlocking relationships.

Tricks for Survival

For people, the Atlantic coast is a pleasant place to swim, explore, or bask in the sun. But for the members of biological communities that inhabit the coast, it is a harsh place to live. They must deal with shifting tides, ocean salt, and pounding waves. As a result, plants and animals of the coast have developed unique adaptations that help them get along. **Adaptations** are special features developed by organisms to help them survive in a particular environment.

The water level along the coast shifts constantly with the tide.

Tides are one of the most important factors in coastal ecosystems. They make for an ever-changing world. The **tide** is the daily rise and fall in sea level along the coast. High tide is the highest point on the shore reached by the sea each day. Low tide is the lowest point. Most coastal areas have two high and two low tides each day.

Many of the creatures in coastal ecosystems live in the **intertidal zone**, the strip of the shore that lies between the highest and lowest tidal levels. At high tide, most creatures find themselves underwater. Then at low tide, many are exposed to the air. So intertidal plants and animals have developed adaptations that allow them to live in both worlds. For example, white acorn barnacles are small intertidal animals that spend their whole lives inside shells stuck to coastal rocks. When the high tide covers them, they open up their shells to take in food from the ocean water. When low tide leaves them exposed, they close off their shells to retain moisture and keep out predators. A **predator** is an animal that hunts or kills other organisms for food.

Salt is another challenge coastal organisms must face. Plants and animals are usually equipped to live in either salt water or freshwater. But the water in some bays and salt marshes is a mixture of the two. So some creatures have developed behaviors and adaptations that help them deal with sudden changes in the **salinity**, or salt content, of the water. Oysters close their shells when the salt concentration in the water around them becomes too high. Salmon, which live in salt water but return to freshwater to lay their eggs, use their kidneys and special glands in their gills to eliminate excess salt from their bodies. Similarly, some salt marsh plants, such as cordgrass, have salt-eliminating cells in their leaves. They can live in salt water that would kill other plants.

Animals and plants on exposed shores have also adapted to the constant pounding of the ocean's waves. Animals on rocky shores, such as barnacles, can stick themselves tightly to rocks, pilings, driftwood, or the hulls of ships. They create a special cement that dries even underwater and allows them to attach to practically anything, including the sides of whales. Those that live on sandy shores burrow into the sand to escape the powerful surf.

Competition

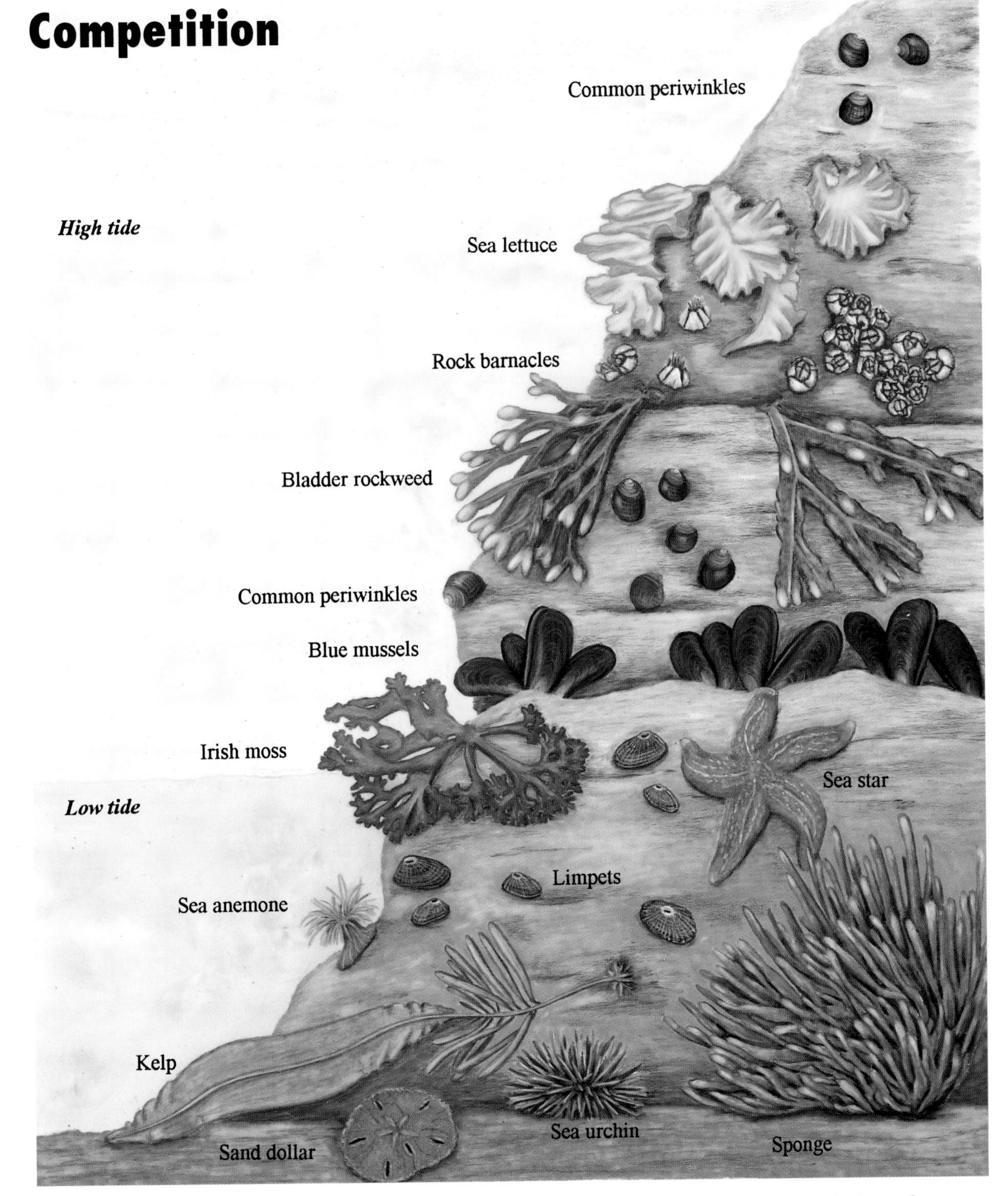

On rocky seashores, organisms not only compete for food and shelter but for places to attach themselves.

Pounding waves make life difficult for the plants and animals of the rocky coast.

A Tough Neighborhood

Adaptations are essential for survival. But each ecosystem has a limited amount of food, water, space, light, and other resources for the organisms that live within it. As a result, **competition**, or the struggle among organisms to get what they need to live, is strong. The competition among members of a community can determine which plants and animals thrive and which do not.

Competition can occur within a species. A **species** is a group of organisms that closely resemble each other and can interbreed. For example, thousands of barnacles may compete with one another for space on a rocky ledge. Competition might also occur between two different species. The higher areas of the rocky shore where the barnacles live are sought by few creatures because they are too dry.

So barnacles have much of that space to themselves. But the area below the barnacles is wetter and populated by many organisms. Barnacles that tried to live there would be crowded out by seaweed and other animals that compete for the same space.

Plants also compete with one another. Certain adaptations allow some plant species to compete more successfully in the intertidal zone. For example, smooth cordgrass grows in salt marshes along much of the Atlantic coast. It is a **halophyte**, a plant that is adapted to live in salt water. Halophytes do well in intertidal areas where salt water soaks their roots for at least an hour or two each day. Because the conditions would be harmful to most plants, halophytes have little competition in this area. But farther away from the shore, where ocean water does not reach, the competition from other plants would crowd out the halophytes.

The Atlantic coast is a unique, often harsh place. The organisms that live there must meet the challenges of both land and sea. As you explore this coast in the next few chapters, you will visit several communities that make up this ecosystem. You will learn how the organisms in these communities live together—vying for position and competing with one another for survival. You will also discover many of the adaptations the plants and animals have been forced to make in order to thrive in their habitats.

This exploration would not be complete without the millions of people who share the Atlantic coast with these plants and animals. How do people compete with them for coastal habitat? What effects do people have? Are some coastal communities threatened by human activity? If so, what are we doing to solve the problem?

Between the Tides

The Atlantic's rocky coastline stretches from the Maritime Provinces of Canada to Massachusetts. Cliffs and boulders face the sea. The wind and waves can be brutal here and make life especially difficult. The creatures that live on this rocky coast are pounded day and night. So what kind of organisms are hearty enough to survive here?

Acadia National Park is a good place to find out. The park is a 40,000-acre (16,000-ha) preserve of rocky shores and evergreen forests, mainly on Maine's Mount Desert Island. When you set out to explore this coast the morning is cool and misty. The forest floor is covered with a soft carpet of moss, decaying leaves, and pine needles. As you near the coast, the sound of the waves grows louder until suddenly, the forest ends and you stand at the edge of the Atlantic.

As you head toward the water, you realize it is low tide. This is the best time to explore because much of the intertidal zone is exposed. You can see many of its plants and animals on or among the rocks.

A dense evergreen forest edges very close to the rocky coastline of Mount Desert Island off Maine's coast.

Looking around, you notice several things right away. First, plants and animals seem to group themselves into horizontal bands on the rock. Animals in lower bands near the ocean need more moisture than those farther away from the water. You can see some of them on the rock. There is a black band near the top (blue-green algae), a white band not far below that (acorn barnacles), and an area of seaweed-covered rock (rockweed) nearer the water.

Some intertidal animals survive in pools left behind at low tide.

You also notice the great competition here for space. Plants and animals seem to cover every inch of space—stuck onto rocks, wedged into cracks. They grow next to and even on top of each other.

The sea is calm now. But you might wonder what happens when there are rough waves. Why don't the tides and the waves just sweep the creatures here off the rocks? The answer is adaptation. Any creature that makes its home on the rocky shore must have a way to grab the rock surface and hold on tight. Sea snails such as periwinkles use a soft, sticky foot to cling to rocks as they slowly creep along. Mussels produce strong, thin threads that attach them to the rocks. Even plants have ways of holding on. Rockweed has holdfasts, rootlike bases that use suction to cling to rocks. Everything on the rocky shore has found a way to keep its seat.

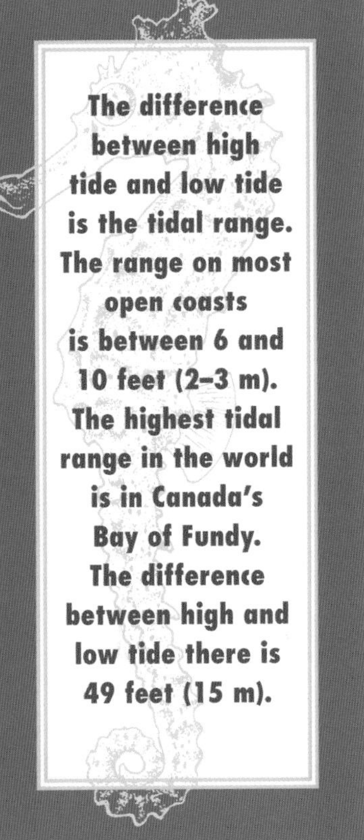
The difference between high tide and low tide is the tidal range. The range on most open coasts is between 6 and 10 feet (2–3 m). The highest tidal range in the world is in Canada's Bay of Fundy. The difference between high and low tide there is 49 feet (15 m).

Where's Your Zone?

Plants and animals that live closest to the sea, in the low intertidal zone, are covered by water most of the time. Only the lowest of the low tides uncovers them. Plants and animals that live farthest from the sea, on the top of the intertidal zone high up on the rocks, are often exposed to the air 90 percent of the time. These creatures must be able to survive summer heat, winter cold, drying winds, and the lack of water.

As a result, creatures that live at the top of the intertidal zone are very different from those that live at the bottom. Lichens and blue-green algae form a black band high up on the rocks in what is called the **splash zone**. It is the top band of the rocks, which is wet only now and then by a splash from large waves or by rain. The algae have a jellylike coating that keeps them from drying out in the sun. In cracks in the rocks just below the band of algae, you see tiny snails called rough periwinkles. They are basically land creatures. They can go for a month without being touched by water. Green crabs live at the other end of the intertidal zone. They are

Periwinkles live high on the rocky coast where waves seldom reach.

true sea creatures, lurking in the low intertidal waters and beyond, where the sea covers them almost all the time.

Competition for space and for protection against predators is also a factor in the location of species on the rocky shore. Barnacles form a white band in the high intertidal zone because it is not a very desirable spot for other organisms. In the low intertidal zone, barnacles would be easy prey for meat-eating snails, such as the dog whelk, which live in greater numbers in the low intertidal zone. **Prey** is an animal that is eaten by another organism.

Mussels live in the midrange of the intertidal zone. They are not found higher in the zone because they cannot tolerate long periods out of water. They are not found lower because they would be eaten by predators, such as the starfish, that live in deeper water.

Even plants fall into zones determined by competition and their individual needs. Rockweed must be submerged for about one hour during each tide. A higher location would dry it out, and it would eventually die. It finds the right conditions on middle intertidal rocks. It cannot take hold lower in the intertidal zone because

Rubbery rockweed provides shelter for mussels, crabs, and other creatures at low tide.

it would have to compete with other types of seaweed that are better adapted to life in deeper water.

Eat or Be Eaten

The rocky shore looks quiet, especially at low tide. But it is the scene of a constant battle of predator against prey. Some animals, such as barnacles, clams, and mussels, are cemented to the rocks. Because they are stationary and have very few defenses, low tide is the safest time for them. Sure, they sit high and dry. But they close their shells to avoid the heat or cold, to conserve moisture, and to save themselves from land predators such as the seagull. Many sea predators live in the lower intertidal zone and in the deeper waters

Barnacles use their feathery feet to sift food from the water at high tide.

beyond it. At low tide, they are forced farther offshore, out of reach of their prey.

Most activity takes place in the intertidal zone at high tide, and mostly at night. That is when many of the creatures of the rocky shore feed. The tidal waters bring algae and **plankton**, microscopic plants and animals that float in the water. Passive feeders (organisms that stay in one spot catching food as it floats by them), such as clams and mussels, open their shells, ready to filter their meals of plankton out of the waters. When the waters of the high tide wash over them, barnacles open the trapdoors at the tops of their shells. They then extend six pairs of slender, delicate legs, using them to pluck food from the water and draw it into their mouths.

Tiny sea snails, such as the limpet and the rock periwinkle, move slowly across the rock to get their meals. The limpet, which lives high in the intertidal zone, stays relatively inactive during the dry times at low tide. But when high-tide waters creep up the rocks, the limpet begins to crawl slowly over the rock, scraping off algae with the sharp teeth on its tonguelike radula.

High tide is also the most active time for meat-eating predators in the intertidal zone. As the water level rises, they can swim farther up the rocky shore in search of prey. Dog whelks (sea snails), starfish, crabs, and sea urchins move in. Even a closed shell is often no defense for a clam, barnacle, or mussel. A starfish can easily use its many arms to force open a clam shell. The starfish then inserts its stomach into the clam's shell to digest the soft animal inside. Dog whelks force open the trapdoor at the top of barnacle shells. They also drill into the shells of mussels with their radula, in order to suck out their meal. Crabs can crack open a clamshell with powerful claws.

Watch Your Step

Rocky coasts are popular with people, too. At low tide, they scramble over pebbly flats, look under rocks, and explore tidal pools. Unfortunately, too many people walk on some rocky shores. They trample on the creatures that live there or collect so many that they upset the balance of the ecosystem.

When exploring the rocky coast, avoid disturbing the plants and animals that live there.

If you visit a rocky coast, watch your step. It is easy to crush small worms, barnacles, crabs, and marine snails that live on rocks or under seaweed. If you lift up rocks to see animals living beneath them, replace them so the animals are not disturbed. Observe the animals, but avoid collecting them. You can appreciate and enjoy the rocky coast, while also helping to preserve it.

Exploring Why Ocean Waters Do Not Usually Freeze

In general, deep ocean waters stay ice-free when lakes, ponds, and rivers inland are covered with ice. Using just a couple of simple ingredients, take a closer look at why ocean waters stay ice-free longer than other bodies of water. You will need:

- **2 medium-size paper cups**
- **1 tablespoon of salt**
- **a spoon**
- **a crayon**

1. **Fill one half of each cup with tap water.**

2. **Put the salt into one of the cups and stir until it has dissolved.**

3. **Write an "S" on the cup with the salt. Leave the other cup blank.**

4. **Put both cups in the freezer. During the first day, check the cups every half hour for about 6 hours. Record what you observe.**

5. **Then leave the cups in the freezer without checking for another day.**

6. **Check again. Record what you observe.**

The water in the "S" cup will not freeze because it contains salt. The salt lowers the freezing temperature of the water. This is the main reason that in areas where lakes and ponds might freeze, coastal waters remain ice-free. In arctic areas, the temperatures stay low long enough to freeze the surface areas of the Arctic Ocean in midwinter.

Shifting Shores

The part of the Atlantic coast characterized by wide, sandy beaches stretches for hundreds of miles from New York to Florida. Unlike the rocky shore, which seemed crowded with life, the sandy beach looks empty. But this is deceptive. The sandy beach is full of living creatures, too. Only they are mostly underground.

Creatures of the sandy beach's intertidal zone have gone underground for good reason. First, they need the same kind of protection from the pounding surf as the creatures of the rocky shore. But on a sandy beach, there are no rocks to grab onto. The sand is loose and shifting. Waves move it around. So creatures such as worms, clams, and crabs live in small holes and tunnels under the sand for protection from the ocean and the weather. They also stay under the sand for protection from daytime predators such as gulls.

Farther up the beach, away from the water's edge, sand dunes hide other creatures. The dunes form when windblown sand is trapped by plants, debris, or fences. Hardy plants such as dusty miller and beach grass hold the dunes in place with their roots. These plants are adapted to withstand the hot sun and dry winds away from the reach of the waves. The dunes also provide food and shelter for creatures such as birds and crabs that often visit the water's edge for food.

The sandy coast looks empty but is full of life.

Cape Hatteras National Seashore in North Carolina is a good place to explore the biological community of a sandy beach. Hatteras Island is part of the Outer Banks, a thin strip of barrier islands that arch out into the Atlantic from the mainland of North Carolina and then turn landward again. A **barrier island** is a low, narrow, sandy island situated parallel to the mainland shore. The beach is wide and flat. It slopes gently up from the ocean to a low line of sand dunes covered with tall spikes of grass. The world's longest chain of barrier islands stretches for 1,500 miles along the Atlantic's sandy beaches from Long Island, New York, to the tip of Florida.

Ghost crabs live on land but return to the sea several times each day.

Come Out Wherever You Are

If you arrive in the middle of the day, you will see lots of sand, but not much action. That is when many of the beach's animal residents stay out of sight. But you can still see evidence of them—if you know where to look.

To find the hideouts of the intertidal creatures, look first in areas of wet sand, especially where the waters are calm. You might see small holes that bubble as you approach or when a wave passes over them. The holes give away the hiding places of creatures such as clams and mole crabs. You might also find small tubes poking up through the sand. Clams and mussels use them to draw in food from seawater washing over the sand. Small mounds of sand with a tiny hole at the center mark the U-shaped burrows of lugworms. All of these animals stay under the sand, out of sight. Farther up the beach, near the dunes, small holes mark the presence of ghost crabs. During the day, ghost crabs stay mostly in their tunnels to escape attacks from birds. But sit still on the beach and you might see some of these tiny creatures scoot quickly across the sand. Although they live on land, they cannot breathe air. So they dash to the edge of the sea several times each day to wet their gills. **Gills** are organs that allow organisms to breathe by removing oxygen from water. A quick drenching from a large wave fills a chamber in the crab's gills with water. The crab then retreats to its burrow with enough oxygen for the next few hours.

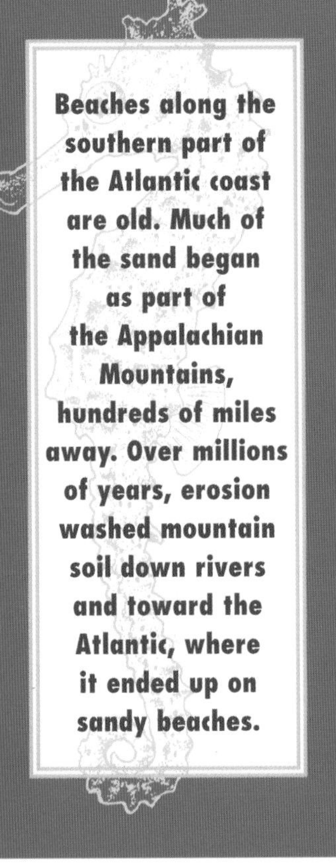

Beach Buffet

Much of the activity of beach animals centers around the competition for food. The waters offshore, just beyond the intertidal zone, are rich in plant and animal life. Plants and animals wash ashore with every wave. This includes plankton by the millions. Microscopic phytoplankton are plants that use photosynthesis to make food, just as larger land plants do. The phytoplankton

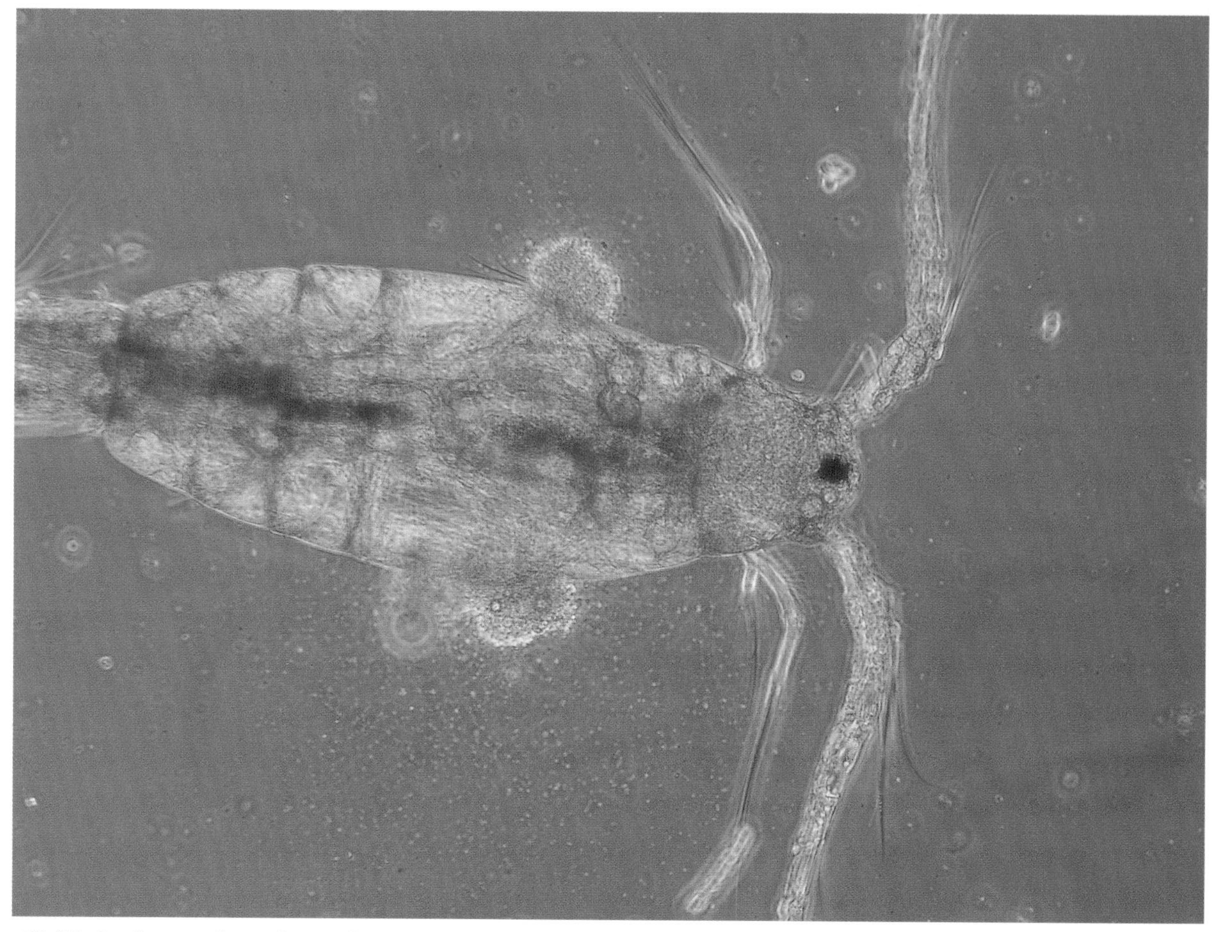

All life in the sea depends on plankton, the base of ocean food chains.

provide food for tiny animals in the water called zooplankton. The plankton are at the base of the ocean food chain. A **food chain** is the linking up of feeding relationships in a biological community. In other words, a food chain tells who eats whom.

Just as on the rocky shore, high tide is feeding time on the beach. Most of the creatures that burrow in the sand are passive feeders. As waves of the high tide wash over the tubes sticking out of their holes and burrows, buried creatures take in water. As they pump water through their bodies, they filter out the plankton and algae suspended in it.

Creatures such as tiny mole crabs are also algae eaters. They live in the wet sand of the intertidal zone and fish for their meals in the surf as the tide comes in. Mole crabs station themselves just beyond where most of the waves are breaking, partly buried with only their feathery antennae and mouth parts sticking out. They trap algae in their antennae as waves that have washed over them slip back into the sea.

Long-necked or soft-shell clams dig holes as deep as one foot (30 cm) in the beach, stretching long siphons to the surface to feed. The clams withdraw their siphons when they feel the vibrations of approaching footsteps. Jets of water sprouting out of the sand often give them away as they quickly pull in these feeding tubes.

Many species of birds compete for food along the Atlantic coast.

Farther up the food chain are meat-eating creatures that prey on plankton and algae eaters such as clams and mole crabs. The knobbed whelk is a meat-eating snail. It uses its powerful foot to hold clams or oysters while it forces the edge of its shell between the two halves of the clam's shell. It forces the shell open the same way you might use your fingernail to pop open the shell of a pistachio nut.

Mole crabs are meals for many other meat-eating creatures, including blue crabs that skitter out of the surf at high tide. Predatory fish move in at high tide, too—just as they do on the rocky shore.

Birds such as herring gulls swoop down to grab whelks and clams. They cannot bite through closed shells. But they will drop the clams onto the ground from a height of twenty feet (6 m) or so in the air until the clam opens and they can eat the tender meat inside. The beach is also home to other birds, such as the common tern, which dives into the water to catch fish and shrimp. You might see plovers or sandpipers just beyond the waves, pecking at the sand with their bills. They are plucking out tiny creatures such as worms, insects, or mole crabs hidden just below the sand. Birds with longer bills, such as willets, can dig in the sand for creatures that burrow more deeply.

Waves continuously wash sea creatures—both dead and alive—onto the shore where they become dinner for scavengers that live on and in the sand. **Scavengers** are creatures that consume dead organisms and their remains. They find food in the line of debris left by the waves that reach farthest up the shoreline. Along with bits of seaweed, driftwood, and shells, the debris includes dead fish and, in late summer, stranded jellyfish at the end of their seasonal life cycles. During the day, the food in this sea rubbish makes a great buffet for scavenging shorebirds such as fish crows.

The Atlantic Ocean near Cape Hatteras is called the "Graveyard of the Atlantic" because so many ships have sunk there. Treacherous currents and frequent storms make this one of the most dangerous areas of the coast. Cape Hatteras Lighthouse at 20 stories is the nation's tallest lighthouse. Since 1870, it has warned ships away from danger.

Discovering Hidden Animals of the Sandy Beach

Beach sand looks lifeless, but it's not. Use a microscope to find out what is really lurking between those grains of sand. You will need:

- **a cup of beach sand**
- **a cup of seawater**
- **a microscope**
- **two microscope slides**
- **two cover slips**
- **an eyedropper or pipette**
- **a notebook**
- **a pencil**

1. **Look at a few grains of sand under the microscope. Can you see small animals?**

2. **Now put a drop of seawater on the slide with the dropper or pipette. Look at this under the microscope as well. What do you see?**

3. **Put a handful of sand into the seawater. Wait a minute or so, then use the dropper or pipette to place a drop of this water on the slide.**

4. **Look at the drop of seawater under the microscope. What do you see now?**

Development has displaced many plants and animals of the sandy coast.

At night along beaches that border forests, raccoons, skunks, and many red foxes may visit the tide line. They are in search of small clams, crabs, or dead fish washed up on shore. They are joined by ghost crabs that leave their burrows in the dunes and prowl about, much safer under the cover of the dark.

Make Way on the Beach

Chances are you have not seen many of the creatures described above on the beach. First of all, most of them come out at night. But another reason is that there are fewer and fewer beaches where mole crabs, ghost crabs, clams, sand hoppers, lugworms, and the thousands of other creatures of the sandy beach community can live undisturbed.

The Atlantic coast is one of the most densely populated parts of the United States. As soon as the weather is warm, people flock to the shore to sun and swim. The scene on most sandy beaches is not empty sand and a few tiny beach creatures poking out now and then. It is boardwalks, beach umbrellas, and hot dog stands. An increasing number of houses perch at the edge of the sand as well.

Coastal wetlands and sandy shorelines have been bulldozed, paved, and built over for resorts, shopping centers, and housing developments. In some places, practically every trace of the natural shoreline has disappeared. In this competition for space, animals and plants of the sandy beach community often lose their habitat and sources of food. Luckily, the federal government and many state governments have taken steps to preserve miles of sandy beaches as state parks, national seashores, wildlife refuges, and recreation areas along the Atlantic coast. One reason this was possible is that many of the Atlantic's sandy beaches are on islands and peninsulas. For long periods of time, there were no roads and bridges to these shores, so they were difficult for many people to reach. Parts of the sandy coast stayed undeveloped long enough for government agencies to step in and preserve the land. In these places, such as Cape Hatteras National Seashore, people can enjoy the sandy beach in its natural state, while observing rules that help preserve the homes and food sources of the many plants and animals of the sandy beach community.

The Ocean's Nursery

Chesapeake Bay is tucked between the western part of Maryland and the Delmarva Peninsula (a strip of land made up of Delaware and parts of Maryland and Virginia). The Chesapeake is an **estuary**, a partly enclosed arm of the sea in which freshwater from rivers and salt water from the sea mix. Two hundred miles (124 km) long from north to south, the Chesapeake Bay is the largest estuary along the Atlantic coast.

Like all estuaries, the Chesapeake is a mixture of many types of natural environments—salt marsh, mud flat, and open water. The fertile mud and marsh grasses supply food to many types of animals, including fish, crabs, and seabirds. The wealth of animals has, in turn, attracted people to the shore of the Chesapeake, too.

People have lived around the bay for centuries, starting with the Algonquin Indians, who arrived here thousands of years ago. Europeans came next, establishing some of North America's earliest English settlements along the Chesapeake's shores. The descendants of some of the Chesapeake's earliest European settlers still live on islands of the bay. These "watermen" have made a living by pulling fish, crabs, and oysters from its waters for hundreds of years.

The fertile Chesapeake Bay provides most of the oysters caught in the United States.

The pressures that people have put on the bay by using it for fishing, crabbing, drinking water, recreation, and dumping have had a negative effect on the Chesapeake's biological communities. But the plants and animals of an estuary community are tough creatures. They have to be tough to survive.

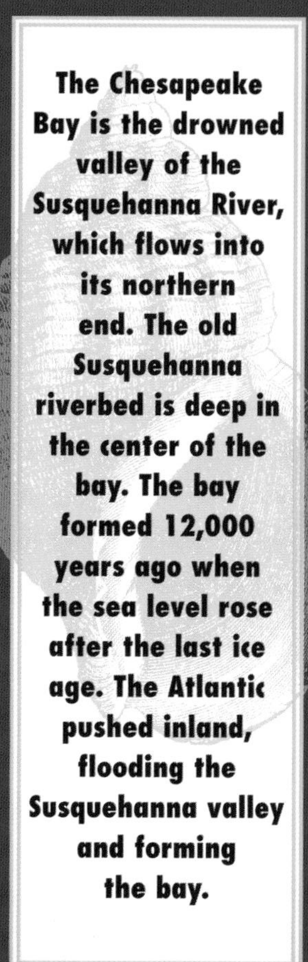

The Chesapeake Bay is the drowned valley of the Susquehanna River, which flows into its northern end. The old Susquehanna riverbed is deep in the center of the bay. The bay formed 12,000 years ago when the sea level rose after the last ice age. The Atlantic pushed inland, flooding the Susquehanna valley and forming the bay.

Pass the Salt

Salt. It is the biggest problem facing animals of the biological communities of estuaries such as the Chesapeake. The salinity of the estuary shifts several times each day with the tides. At high tide, salty seawater pushes in from the sea, flowing up the estuary. At low tide, freshwater from rivers flows into the estuary as seawater recedes. So the water in the bay has varying amounts of salt in it, depending on the time of day, as well as water depth.

Salty seawater is denser than freshwater from rivers. So seawater enters the estuary along its bottom, with freshwater spreading through the estuary along the surface. As a result, salinity in most estuaries not only varies depending on the distance from the sea, but also on the distance from the surface to the bottom. Salt content increases the deeper you go and also the closer you get to the estuary's mouth.

Plants and animals deal with these daily shifts in salinity in several ways. In general, many plants and animals simply live in parts of the bay where the salinity suits them best. So you might find starfish at the bay's mouth, where salty seawater is the rule. At the upper end of the bay, where there is hardly a trace of salt, freshwater fish species such as American shad are common.

But shifts in salinity are not always predictable. As a result, many organisms of estuary communities must be able to adapt to frequent changes. If an animal cannot adjust to these constantly shifting levels of salinity, it cannot survive as part of the estuary's biological community. **Bivalves** (sea creatures with a shell made

Starfish use their powerful arms to pry open the shells of animals such as clams.

of two hinged halves, such as oysters, mussels, and clams) just close up. They can close their shells tightly when the salt content is not right for them. Many animals cope by eliminating excess salt from their bodies. Others increase the internal pressure of their bodies to equal that of the water outside, keeping water of an undesirable salinity out. Other animals, such as crabs, have skin or shells that water cannot penetrate.

Fertile Waters

Estuaries and the grassy wetlands that border them are the habitat of a great concentration of sea creatures. A wetland is a land area that is covered most of the time by water. Rivers entering the

Chesapeake Bay carry sediment, which is deposited along its margins. Over time, mudflats build up at its edges. At low tide, the remains of small sea animals and plants are left to dry in the sun. As they decay, they fertilize the sediments, making them a good place for grasses to take root. The Chesapeake is very rich in plant and animal life. A major reason is that decaying plant matter from the grasses in and around the estuary provide food for a type of plankton called phytoplankton. **Phytoplankton** are tiny floating plants that are the first link at the base of aquatic food chains.

Mudflats act as a crossroads for a wide variety of creatures. Animals from the forest, marsh, and estuary are drawn to the shore.

The plankton, algae, and submerged grasses provide a banquet for young fish, crabs, clams, shrimp, oysters, and other marine life. The shallows and grassy margins of the Chesapeake act as a nursery for many waterfowl that nest on its margins. Blue crabs that shed their hard shells as they grow hide in the submerged grasses as their new soft shells harden. Predatory crabs, fish, and birds also hunt there for food.

As a result of all this traffic, the number of fish and other living creatures in an estuary tends to be much greater than the number of creatures in another body of water of similar size. As many as 70 percent of the fish that swim in the waters off the Atlantic coast spawn in or spend part of their lives in the waters of the Chesapeake or the rivers that flow into it. Estuaries are often the great nurseries of the ocean.

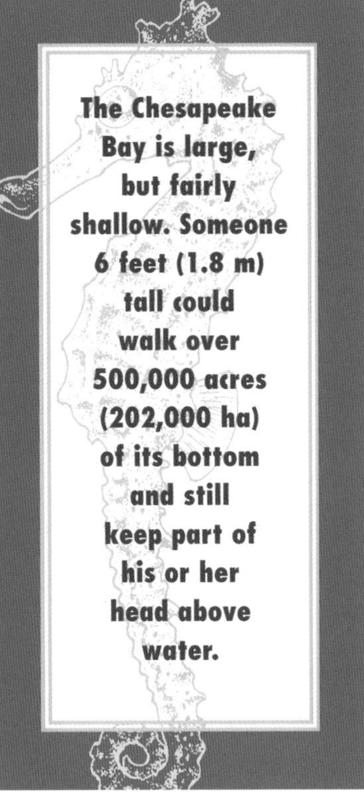
The Chesapeake Bay is large, but fairly shallow. Someone 6 feet (1.8 m) tall could walk over 500,000 acres (202,000 ha) of its bottom and still keep part of his or her head above water.

Catch of the Day

Many biological communities share the Chesapeake. One of the most interesting are the oysters, which are an important part of the Chesapeake Bay ecosystem. They settle down together, forming reefs or bars on the bottom of the bay, which provide a habitat for many other estuary species. Grass shrimp, anemones, barnacles, hooked mussels, mud crabs, and red beard sponges can also be found there. Oysters serve as food for many species, including people. What is more, they are also little filters, taking pollution, grit, and floating organic material out of the bay's waters.

At the beginning of their life cycle, oyster larvae attach themselves to a rough, hard surface, such as another oyster shell, at the bottom of the bay. As they mature, they secrete material to produce a shell of their own that protects the soft animal inside. The adult oysters, which produce shells that can grow to 15 inches (38 cm), crowd together, growing in clumps that form bars on the bay's bottom.

Thus, predators that feed on oysters, such as starfish and small snail-like creatures called oyster drills, also live at the oyster bars.

Other larger predators, such as striped bass, weakfish, croakers, and blue crabs, frequent the oyster bars because they know they will find many other animals there that can be eaten. This in turn makes the bars a favorite spot for fishers and crabbers.

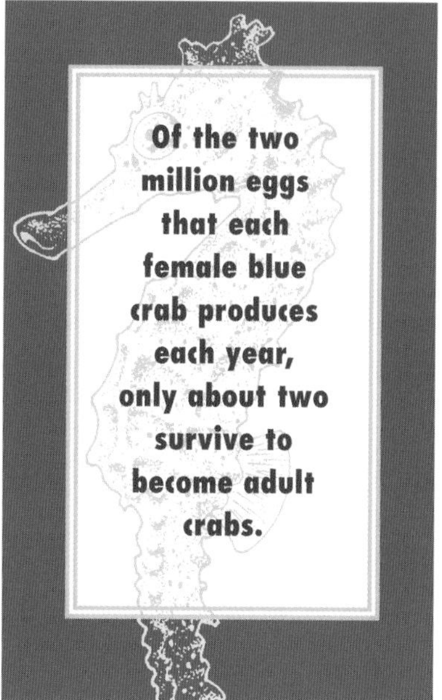

Adaptation and competition combine to determine the range of bay oysters and oyster predators. For example, predatory oyster drills and starfish thrive in very salty water at the mouth of the estuary. So oysters there are under constant attack. Luckily for oysters, the starfish cannot stand much of a decrease in salinity. So oysters can rid themselves of these pests by moving slightly inside the mouth of the bay, where the salt content is a bit lower.

Oyster drills can tolerate this slight decrease in salinity and will pursue the oysters inside the bay's mouth. But they will not go where the bay is just one-third salt water. The salinity there is not the best for oysters either. They need a certain amount of salt water or they will die. So oysters that live in this part of the bay do not grow very large. But because oysters can live there free of predators, some beds do exist well inside the bay.

Bay Watch

Oysters were the main catch of the bay until the 1980s, when their numbers decreased as a result of overharvesting. In the 1920s, catches totaled more than 50 million pounds (22.7 million kg) each year. Now they are down to less than 5 million pounds (2.27 million kg) per year. The oyster population in the bay is now about 1 percent of what it was just over one hundred years ago.

In the mid-1980s, watermen turned to blue crabs to make up for the decreased numbers of oysters. Blue crabs spend most of their lives in the bay, settling down into the bottom mud to hibernate during each winter. The crabs use their six walking legs to scuttle along the bay's bottom. Their two paddlelike back legs help them swim. Powerful front claws are used for eating and defense. The crabs are tough creatures, and they have thrived in the bay. But the increased attention to crabbing has caused a decline in the blue crab population as well. Almost 150 million pounds of crabs are caught in the bay each year—more than half of all the blue crabs caught in the United States.

Although the decline in blue crabs is troubling, the decline of oysters has affected the Chesapeake in one very important way. Fewer oysters are able to filter less water. The bay once held enough oysters to filter its water in three to six days. The oysters in the bay today would take more than a year to do so.

But the news is not all bad. People are taking steps to solve the bay's problems. The government has placed limits on the number of oysters, crabs, and fish that can be caught in the bay to give populations time to increase. Laws to decrease water pollution and conserve wetlands have also helped clean the bay's waters and preserve this important habitat.

Sea of Grass

Enter a salt marsh and all you are likely to see at first is grass growing everywhere you look. You might wonder what could be so special about such a place. The answer is not in how a salt marsh looks. It is what a salt marsh does.

A **salt marsh** is a grassy wetland, washed a couple of times each day by salty tidal waters. It is a buffer zone that protects coastal areas behind it from storm damage and the constant wash of waves. It is also a filter that removes pollution from coastal waters. It serves as a nursery for young fish, birds, shrimp, and crabs. It is also home to and the feeding grounds for many species of plants and animals. Like the oyster bars of Chesapeake Bay, a salt marsh is the crossroads of a vast number of organisms.

Along the Atlantic coast, you will find salt marshes in the sheltered areas of estuaries such as Chesapeake Bay and on the landward sides of barrier islands. The salt marsh that we will explore here is located on the landward side of Cumberland Island, a barrier island off the coast of Georgia. Because of its unique environment, the federal government decided to preserve this island by creating Cumberland Island National Seashore in 1972.

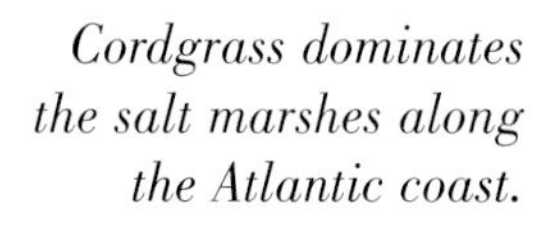

Cordgrass dominates the salt marshes along the Atlantic coast.

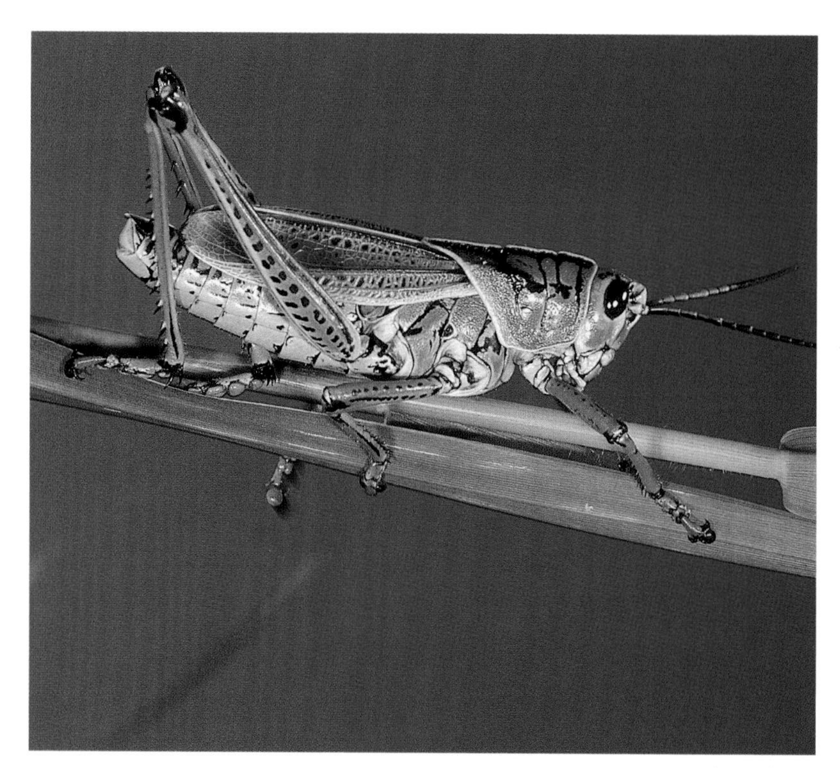

The grasses of the salt marsh provide food and shelter for insects, birds, and other animals.

Grass and More Grass

In a salt marsh, the tall grasses stretch as far as the eye can see. Along the southern Atlantic coast, smooth cordgrass dominates. These tough, broad-leafed grasses can grow more than six feet (1.8 m) tall. In summer cordgrass is green; but in the fall, it turns brilliant gold.

Cordgrass forms the foundation of the marsh. Its roots grow horizontally in the ground. New grass stems shoot up from the roots growing sideways, just under the surface of the land. The root structure holds marsh soil together and helps prevent erosion.

The grasses thrive on the low-lying western edge of Cumberland Island, where seawater invades twice a day at high tide. Cordgrass flourishes where other plants cannot live because it can remove salt from most of the water that enters its system. It keeps dissolved salts in its roots and extracts it using special cells on its leaves. You can sometimes see the extracted salt as white patches on its leaves.

The Key to the Marsh

Marsh creatures are very dependent on cordgrass. Insects such as grasshoppers and marine animals such as snails eat a small percentage of the grass. But the rest of the cordgrass forms the basis of salt marsh food chains.

When cordgrass dies, dry, brittle pieces break off and fall into the waters of the marsh. There bacteria and fungi break it down into nutrients that feed plankton. Mixed with bacteria and algae, it also feeds flies and snails. Carried by marsh waters, it is food for filter feeders such as clams, mussels, and oysters.

Cordgrass also creates food and habitat for other marsh animals. Small marsh periwinkles cling to its stems and crawl in the mud searching for decaying organic matter. Small, black mud snails feed at the base of the tough stalks.

At high tide, invading seawater brings predators to the intertidal zone—just as on the rocky shore and the sandy beach. Blue crabs and fish such as the striped mullet move farther up marsh creeks hunting for prey.

The diamondback terrapin thrives at the top of the salt marsh food chain.

At low tide in the marsh, you can find wading birds, such as wood storks, great egrets, and clapper rails, searching for food in the tidal creeks at the edge of grass fields. Overhead, pelicans and gulls scan the ground for food. Ospreys dive into nearby waters snatching fish in their talons. Diamondback terrapin turtles and brown fiddler crabs leave their hiding places to search for food. All of these creatures and everything they eat are nourished by the salt marsh's sea of grass.

Marshlands have been filled in to create new homes, shopping centers, and offices.

Eye on the Future

There are fewer coastal wetlands today than there were many years ago. Long ago, wetlands were seen as soggy swamps filled with pesky mosquitoes. American colonists began filling them in. The Swamp Land Act of 1850 encouraged the draining and filling of wetlands for more "productive" uses. Unfortunately, the plan worked. Since then, about half the total wetland acreage of the coastal United States has been turned into dry land for homes, factories, roads, and shopping centers, all in the line of coastal storms and runoff.

This situation is changing as we realize the value of wetlands as a habitat for many types of plants and animals, as a buffer to protect the shore, and as a sponge to relieve flooding and remove pollution from waters flowing through them.

The federal government and several states now have a "no net loss" policy toward wetlands. This policy states that wetlands should not be destroyed for other land uses. But if any new development does take place in wetlands, the area of wetland destroyed must be matched by restoring or creating an equal or greater amount of new wetlands. Still, it may take decades for these re-created wetlands to approach the productivity level of the original wetlands.

Rain Forest of the Sea

The last ecosystem you will visit is one of the most unusual. You will have to take a boat to travel to this underwater world of bright purples, fiery oranges, deep greens, and brilliant yellows.

Coral reefs are underwater structures formed from the limestone skeletons of billions of tiny animals called coral polyps. In and around them are thousands of species of colorful animals and unusual plants. In fact, coral reefs are among the richest biological communities on Earth. Much like a rain forest, one square yard (.8 sq m) of coral reef can contain tens of thousands of individual corals and more than a hundred different species—fish, crabs, clams, sponges, and plants.

Coral reefs form only in the warm, shallow waters of the world's midlatitudes. The U.S. mainland has one chain of coral reefs off the Atlantic coast of Florida. It stretches from Miami down to a string of islands at the southern tip of Florida. These islands are called the Keys. The Keys themselves are the tops of old coral reefs. Now a new chain is being formed between one and three miles to the east of the Keys.

Coral reefs develop in shallow, warm, clear waters.

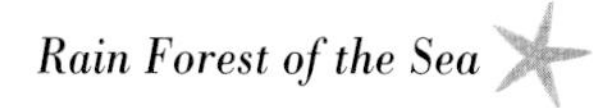

Billions of Tiny Animals

The outer layer of the reef is alive with colonies of living coral polyps growing on the old reef underneath. Each polyp is a small, soft, tubelike animal about the size of your thumbnail. It has a central mouth ringed with tentacles. When young, polyps use the minerals in seawater to create limestone shells in which they encase themselves. The coral spends its life in that case, sticking its tentacles out only to catch plankton floating by in the seawater. When polyps die, their shells become part of the reef, and new living corals grow on top of them. In this way reef building can go on for thousands of years.

Undersea earthquakes can send walls of water called tsunamis speeding toward coasts at up to 500 miles (800 km) per hour. When they reach the coast, tsunamis can be 60 feet (20 m) high. A 1960 earthquake off the coast of Chile sent a tsunami racing across thousands of miles of ocean to wash over part of the Japanese coast. Luckily, such tsunamis are rare.

The shells of millions of tiny coral polyps form huge coral reefs.

The coral reef ecosystem is the most diverse of the ocean environment.

The corals live in symbiosis with a tiny, colorful type of one-celled algae. **Symbiosis** is a dependent relationship between organisms from which both benefit. The algae are tiny plants that live inside the polyps, giving the corals their color. The algae provide food and oxygen for the polyps through photosynthesis. In turn, the algae use the waste that coral polyps produce as food. Because the algae need sunlight, coral reefs grow only in clear, shallow waters.

Different coral species grow in a number of shapes—antlers, piles of twigs, large brains, trees, fans, and boulders. The overall shape of the coral reef depends on the species of coral that form it. Boulder coral grows in tan or greenish mounds, while elkhorn coral grows in flattened yellow branches.

A Little of This, A Little of That

A great variety of animals and plants also live on the surface of the reef. Sponges and worms stay in one spot on the reef, sucking in water to remove plankton for food. Large predators, such as sharks and barracudas, glide through the water patrolling the area for prey.

Giant clams, starfish, and sea anemones can be found nearby. Octopuses and moray eels hide in dark crevices in the reef, coming out only at night to feed. Colorful schools of fish swim by, darting here and there in magnificent formations.

On the reef, animals live with one another, on top of one another, and sometimes inside one another in close commensal relationships. **Commensalism** is a relationship in which one organism benefits and the other is neither helped nor harmed. Large sponges, with their many holes, provide opportunities for commensal relationships with other creatures such as the worms or shrimp that live in them. The small animals get a safe shelter from predators. They also share the sponge's food. The sponge does not benefit from the relationship, but its survival is not threatened either.

In early America, trade between some Atlantic coast Indians and Europeans was conducted using white and purple beads called wampum. White beads came from the shells of sea snails called whelks. Purple beads came from the hard-shell clam, commonly called quahogs.

Other reef creatures have cooperative symbiotic relationships from which both parties benefit. Some species of fish and shrimp have "cleaning stations" where they clean parasites from other reef animals. A **parasite** is an organism that lives in or on another and harms it in the process. "Cleaner fish" benefit by eating the parasites, and the fish that are cleaned get rid of parasitic organisms. The red-and-white-striped coral shrimp waves its antennae to catch the attention of passing fish and let them know it is in business. The neon goby advertises its cleaning services by swimming back and forth in certain patterns. The fish that want to be cleaned then flip themselves so they are head-down or tail-down to signal that they are ready for cleaning. Cleaner fish even remove parasites from inside the mouths of the larger fish without being eaten.

Not all reef creatures cooperate. The competition on the reef for food and space is also great. Corals eat plankton and algae. Snails and fish eat corals. Sharks and rays eat anything they want.

To escape predators and to reduce the chance of being eaten, some animals use camouflage. The arrow crab is hidden by its color patterns—a striped body with red and blue on its claws blends into the reef. The sponge crab hides by carrying a living sponge on its back.

A fish becomes a meal for a lobster hunting along the reef.

Bleaching has already claimed several strands of this coral.

Reefs in Danger

The reefs have been an important part of the economy of Key West for years. Home to so many species of fish and other sea life, reefs have helped support the local fishing industry. Reefs also attract tourists, who like to dive and snorkel to look at the coral formations and interesting aquatic life.

But Florida's reefs are in trouble. These reefs are located along one of the most densely populated coastlines in the country. Sewage and runoff of pesticides and fertilizers from farms that end up in coastal waters cause pollution that kills coral. Coastal development that tears up and exposes soil creates sediment-laden runoff that enters coastal waters and smothers coral polyps. Reefs are further damaged by boat anchors, divers, storms, and collectors who chip off pieces of coral to sell to tourists.

Many corals also die from diseases such as bleaching. This disease, caused by overly warm water, causes coral to expel their algae. The corals soon die, leaving only their white limestone skeletons.

Many nations are now creating coral reef reserves. They are also trying to educate people about the importance of reefs and the need to protect them. Florida has set aside John Pennekamp Coral Reef State Park to protect part of the reef chain. In such preserves people can see the reef, but must follow strict rules that prevent them from harming it. The federal government has also set aside a few small pieces of reef as reserves. The Florida Keys National Marine Sanctuary has "no take zones," where marine life cannot be harvested. But these zones cover only a small percentage of the reef. All over the world, people who love reefs and realize their value hope that more will be done to preserve these unique ecosystems.

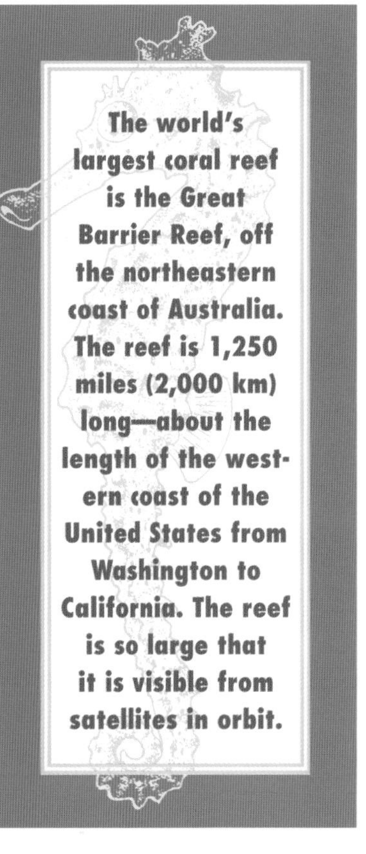

Tomorrow's Coast

People love the coast. Every year, more vacation homes, restaurants, shopping centers, amusement parks, and hotels spring up along the coast. Because most of the ocean's organisms live along its coastal margins, there is bound to be competition between people and coastal biological communities. This is a competition that coastal plants and animals often lose.

Elbow Room

Many of the wide beaches, sand dunes, and marine forests on the coast's barrier islands have been greatly changed by the building of beach resorts and the expansion of coastal towns and cities. Many salt marshes on the edges of coastal islands and estuaries have been filled in to build airports, factories, harbor facilities, and housing developments.

When development gobbles up land on the coast, coastal rocks, beach sand, and cordgrass fields that serve as animal habitat disappear. Animals displaced from their homes cause a disruption in the food chain. Many coastal biological communities can no longer survive.

Development along the coast can destroy the habitats of coastal plants and animals.

The Coastal Sink

The millions of people who live along the Atlantic coast also affect the quality of coastal waters. Runoff sends lawn chemicals into city streets and eventually into coastal waters. Chemicals left over from manufacturing too often end up in coastal rivers and bays. Or they may be dumped miles from the coast. But in time they reach the waters.

The pollution fouls coastal waters, sometimes sickening or killing coastal fish and corals. It also causes the disappearance of coastal grasses that serve as habitat and essential food.

Hope for the Future

The problems of habitat loss and pollution along the coast are serious. So people are taking steps to do something about them.

Much habitat has already been lost. But the federal government and the governments of coastal states have taken steps to stop the complete disappearance of these important habitats and their biological communities. They have created several national and state reserves where coastal ecosystems are preserved in their natural state. These include the rocky coast of Maine's Acadia National

Drains carrying household sewage and industrial waste can pollute coastal waters.

Elevated boardwalks help protect the Everglades environment.

Park, as well as the sandy beaches of Virginia's Assateague National Seashore and North Carolina's Cape Hatteras National Seashore. Lands around the Hudson River, the Delaware River, and the Chesapeake Bay estuaries have been preserved for wildlife, including nesting and migrating birds. In New York City, Gateway National Recreation Area protects the 10,000-acre (4,000-ha) Jamaica Bay Wildlife Refuge from future development. Parts of the Florida coral reefs are further protected as Dry Tortugas National Park.

Federal and state laws to regulate the discharge of sewage and other pollution into coastal waters have controlled this problem in many areas. Although other problems remain, surface waters in many rivers and bays are cleaner now than they have been in several years.

People will continue to live, play, and work on the Atlantic coast. The many biological communities will continue to exist there, even if many of them are now fighting for survival. We now know more about the way plants and animals of the coast live. We also know more about the natural processes and cycles that sustain them. Many people are working to ensure that we use this knowledge to both enjoy the Atlantic coast and preserve its ecosystems.

Glossary

adaptations the special features developed by organisms to help them survive in a particular environment. Leaf cells that remove salt are an adaptation that allows cordgrass to live in seawater.

barrier island a low, narrow, sandy island situated parallel to the mainland shoreline.

biological community all of the organisms that live together and interact in a particular environment. The biological community of the rocky shore includes all of the different plants, animals, and other organisms that live there.

bivalves sea creatures with a shell made of two hinged halves. Oysters, mussels, and clams are bivalves.

commensalism a relationship between organisms in which one organism benefits and the other is neither helped nor harmed. The sponge and the small creatures that live inside its holes, getting shelter and food but not harming the sponge, have a commensal relationship.

competition the struggle among organisms to get what they need for survival.

coral reef an underwater structure formed from the limestone skeletons of coral polyps.

ecosystem the association of living things in a biological community, including their interaction with the nonliving parts of the environment. The biological community of the rocky shore, plus the rocks, ocean, and other nonliving things with which it interacts, make up the rocky shore ecosystem.

estuary a partly enclosed arm of the sea in which freshwater and salt water mix. The Chesapeake Bay is an estuary.

food chain the feeding relationship between organisms in a biological community.

gill an organ that allows creatures to breathe by extracting oxygen from water and releasing carbon dioxide back into it. Fish and many other marine organisms breathe through gills.

habitat a place that has all the living and nonliving things that an organism needs to live and grow. The branches of a tree are a bird's habitat. The sandy bottom of the Chesapeake Bay is the eastern oyster's habitat.

halophyte a plant that can live in salt water.

intertidal zone the part of the shore between the tide's highest and lowest points.

organism a living thing, such as a plant or animal.

parasite an organism that lives in or on another organism and harms it.

phytoplankton microscopic floating plants that are the first link at the base of ocean food chains.

plankton microscopic plants and animals that float in water. Plankton is composed of phytoplankton (tiny plants) and zooplankton (tiny animals).

predator an animal that hunts or kills other animals for food. A starfish that eats mussels is a predator.

prey an animal that is hunted or killed by other animals for food. Mussels are the prey of starfish.

salinity the salt content of a substance, such as water.

salt marsh a grassy wetland, washed a couple of times each day by salty tidal waters.

scavenger a creature that consumes dead organisms and their remains. Seagulls that eat dead fish washed up on a beach are scavengers.

species a group of organisms that closely resemble each other and can interbreed with one another in nature.

splash zone the top band of the rocks along a shoreline, wet only occasionally by splashes from large waves or by precipitation.

symbiosis a dependent relationship between organisms in which both benefit. Fish that are cleaned at cleaning stations on coral reefs have a symbiotic relationship with the animals that remove the parasites from them.

tide the daily rise and fall in sea level along the coast.

tidal zone the strip of the shore that lies between the highest and lowest tidal levels.

Further Exploration

Books

Badger, Curtis. *The Salt Tide*. Mechanicsburg, PA: Stackpole Books, 1993.

Cochrane, Jennifer. *Water Ecology*. New York: The Bookwright Press, 1987.

Doris, Ellen. *Seashore Life*. Danbury, CT: Grolier International, 1996.

Earle, Sylvia A. *Sea Change*. New York: Putnam Press, 1997.

Hecht, Jeff. *Shifting Shore: Rising Seas, Retreating Coastlines*. New York: Charles Scribner's Sons, 1990.

Knapp, Brian. *Beach*. Danbury, CT: Grolier Education Corporation, 1993.

Kochanoff, Peggy. *Beachcombing the Atlantic Coast*. Missoula, MT: Mountain Press, 1997.

Mason, Helen. *Life at the Seashore*. Burlington, Ontario, Canada: Durkin Hayes Publishing, Ltd., 1990.

Sayre, April Pulley. *Seashore*. New York: Twenty-First Century Books, 1996.

Steele, Philip. *Do You Know About Life in the Sea?* New York: Warwick Press, 1986.

Waterlow, Julia. *The Atlantic Ocean*. Austin, TX: Raintree Steck-Vaughn, 1997.

Organizations

Acadia National Park
P.O. Box 177
Bar Harbor, ME 04609
(800) 358-8550

Cape Hatteras National Seashore
Route 1, Box 675
Manteo, NC 27954
(252) 493-2111

Cumberland Island National Seashore
P.O. Box 806
St. Marys, GA 31558
(912) 882-4336

Florida Keys National Marine Sanctuary
216 Ann Street
Key West, FL 33040
(305) 292-0311

John Pennekamp Coral Reef State Park
P.O. Box 487
Key Largo, FL 33037
(305) 451-1202

U.S. Environmental Protection Agency
Chesapeake Bay Program Office
410 Severn Avenue, Suite 109
Annapolis, MD 21403

Index

Page numbers for illustrations are in **boldface**.